nes™

HELICOPTERS

Catherine Ellis

Published in 2007 by The Rosen Publishing Group, Inc.
29 East 21st Street, New York, NY 10010

First Edition

Editor: Amelie von Zumbusch
Book Design: Greg Tucker

Photo Credits: Cover, pp. 7, 9, 15, 17, 19 Shutterstock.com; p. 5 by Master Sgt. Robert W. Valenca, U.S. Air Force; p. 11 Department of Defense photo by SRA Jorge A. Rodriguez, U.S. Air Force; p. 13 Department of Defense photo by PH2 Lisa Aman, U.S. Navy; p. 21 Department of Defense photo by SSGT Myles D. Cullen, U.S. Air Force; p. 23 Department of Defense photo by TSGT James D. Mossman, U.S. Air Force.

Library of Congress Cataloging-in-Publication Data

Ellis, Catherine.
   Helicopters / Catherine Ellis. — 1st ed.
      p. cm. — (Mega military machines)
   Includes index.
   ISBN-13: 978-1-4042-3666-0 (library binding)
   ISBN-10: 1-4042-3666-X (library binding)
   1. Military helicopters—Juvenile literature. I. Title.
   UG1230.E45 2007
   623.74'6047—dc22
                                             2006029632

Manufactured in the United States of America

Contents

People in the military fly helicopters. A helicopter is a kind of aircraft that has a **rotor**.

The **blades** of a helicopter's rotor spin very fast. This is what makes the helicopter fly.

This helicopter is landing. Helicopters do not need much room to land.

The person who flies a helicopter is called a pilot.

The military uses helicopters to fire **missiles**.

Helicopters are also used to pick up people who have been hurt and carry them to a safe place.

This is an Apache helicopter. The U.S. Army has many Apache helicopters.

A Chinook helicopter has two rotors. It can carry heavy things.

Black Hawk helicopters are 65 feet (20 m) long.

This is a Kiowa helicopter. The military has many kinds of helicopters.

Words to Know

blades (BLAYDZ) The long, wide parts of an oar, paddle, or rotor.

missiles (MIH-sulz) Things that are shot at something far away.

pilot (PY-lut) A person who works an aircraft, spacecraft, or large boat.

rotor (ROH-ter) A machine that makes power by turning or spinning.

Index

Web Sites

Due to the changing nature of Internet links, PowerKids Press has developed an online list of Web sites related to this book. This site is updated regularly. Please use this link to access the list:
www.powerkidslinks.com/mmm/helicop/

4/17 (39) 9/16